A GEOLOGIST'S SKETCH BOOK

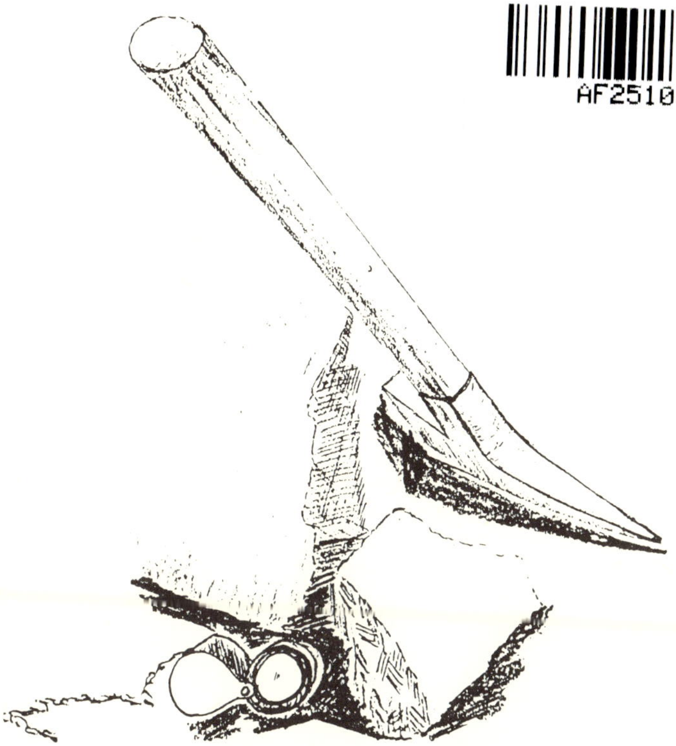

Notes & Drawings
By
C. S. Content

Published by
American Society of Civil Engineers
345 East 47th Street
New York, N.Y. 10017

Price: $8.00

CHARLES S. CONTENT

Charles S. Content is a Consulting Engineering Geologist in San Mateo, Calif. Mr. Content has more than 42 years of experience in engineering geology involving the application of the principles of geology to planning, design, construction and operation of civil engineering structures such as dams, tunnels, canals, bridges, power houses (fossil and nuclear), missile sites, and rapid transit systems.

From 1956 to 1976 Mr. Content was Chief Geologist and Manager of Engineering Geology for the Bechtel Group of Companies. He worked for the U.S. Bureau of Reclamation from 1942 to 1956 as an Assistant Regional Geologist in Sacramento, Calif. and as a Regional Geologist in Billings, Montana. Mr. Content received a B.S. in Geology from the University of Colorado in 1934 and he is a Registered Geologist and Engineering Geologist in California. He is a Fellow in the Geological Society of America, a Member of the Association of Engineering Geologists, and a member of the U.S. Committee on Large Dams.

~INTRODUCTION~

MOST OF THE ITEMS IN THIS SKETCH BOOK PERTAIN TO ENGINEERING GEOLOGY; THE APPLICATION OF THE PRINCIPLES AND PRACTICE OF GEOLOGY TO THE PLANNING, DESIGN AND CONSTRUCTION OF CIVIL ENGINEERING WORKS. BUT THE PETROLEUM GEOLOGIST, THE MINING GEOLOGIST AND THE GENERAL PRACTITIONER SHOULD FIND SOMETHING OF GEOLOGIC INTEREST IN THE BOOK, OR PERHAPS WILL RECOGNIZE A FAMILIAR SCENE.

INTERSPERSED WITH THE GEOLOGICAL INCIDENTS ARE OCCASIONAL SKETCHES OF LANDSCAPES.

C. S. CONTENT.

OLD MT.
PERU
9-72
CONTENT

<u>SUBJECT</u>:   PUMP STORAGE SCHEME

<u>CONDITIONS</u>:   LOWER STORAGE IN ARM OF EXISTING RESERVOIR. UPPER STORAGE ON LAVA TABLELAND.

<u>PROBLEM</u>: BASALT FOUNDATION FOR UPPER STORAGE AREA CONTAINS SCORIACEOUS INTERFLOW ZONES WHICH ARE POTENTIAL PERMEABILITY PATHS FOR WATER LOSS FROM UPPER RESERVOIR.

<u>SOLUTION</u>: BLANKETING WITH LOCALLY AVAILABLE SILTY LOESS. PROJECT UNDER STUDY.

CONTENT '68

<u>SUBJECT</u>: RAILROAD CUT FOR LOOP TRACK TO MINE AND MILL.

<u>CONDITIONS</u>: RAILROAD CUT 300 FEET HIGH IN CONTORTED TALCOSE PHYLLITE.

<u>PROBLEM</u>: OWNER PREFERRED HIGH OPEN CUT RATHER THAN TUNNEL OR TRESTLE TO REACH WORK AREA.

<u>SOLUTION</u>: CUT FAILED DURING CONSTRUCTION DISPLACING SEVERAL THOUSANDS OF CUBIC METERS OF MATERIAL. FAILURE APPEARED TO BE IN LOOSE, FINE GRAINED HEMATITE UNDERLYING PHYLLITE. CONSTRUCTION COMPLETED AFTER REMOVAL OF REQUIRED AMOUNT OF SLIDE DEBRIS.

BRAZIL 9-72
CONTENT

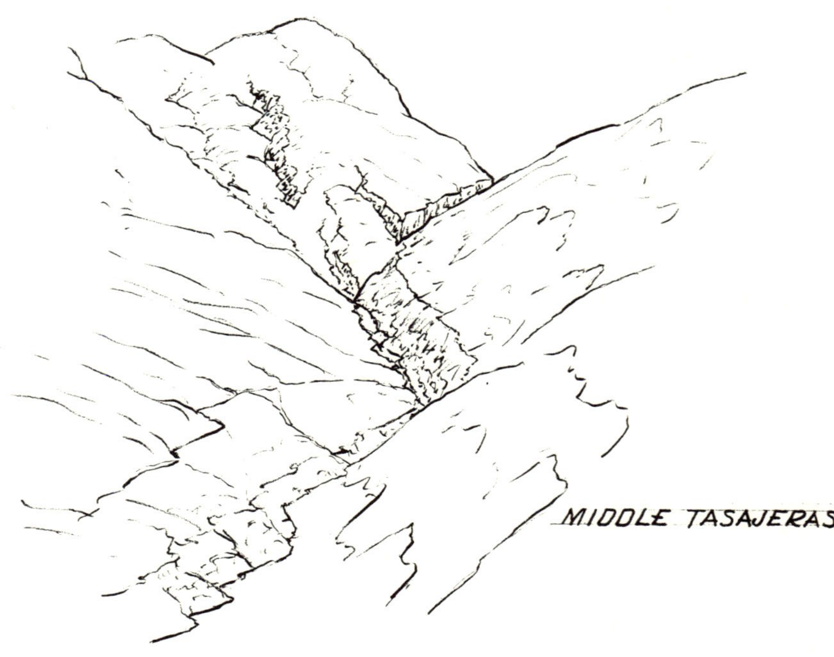
MIDDLE TASAJERAS
PERU 1971
CONTENT

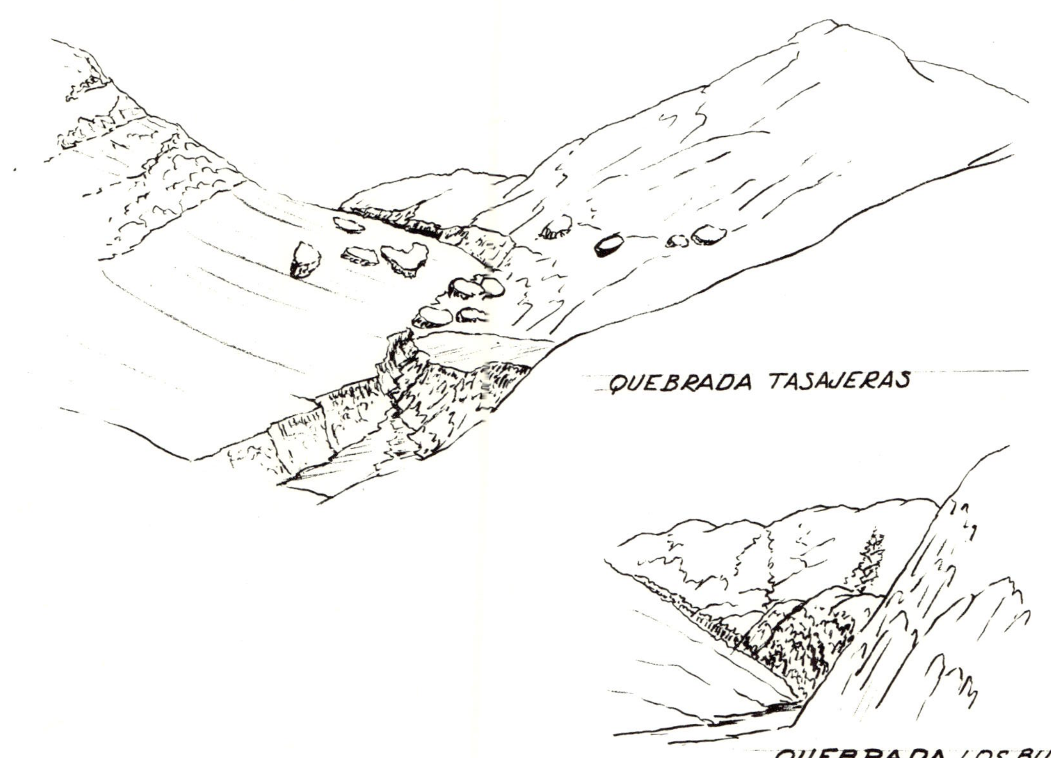

QUEBRADA TASAJERAS
QUEBRADA LOS BURROS
PERU 9-72
CONTENT

SUBJECT: AN EXISTING CONCRETE ARCH DAM

CONDITIONS: FOUNDATION ROCK IS CHIEFLY BRITTLE, WELL-BEDDED RHYOLITE AND CRYSTAL TUFF.

PROBLEM: UNSIGHTLY AND PSYCHOLOGICALLY DISADVANTAGEOUS LEAKAGE SURFACES AT TOE OF DAM ON THE RIGHT ABUTMENT.

SOLUTION: A SINGLE LINE OF GROUT HOLES 10 FEET APART COLLARED 5 FEET ABOVE CONCRETE–ROCK CONTACT. HOLES DRILLED 25 FEET DEEP IN TWO STAGES; FIRST STAGE WITH PACKER 10 FEET OFF BOTTOM, SECOND STAGE AT COLLAR.

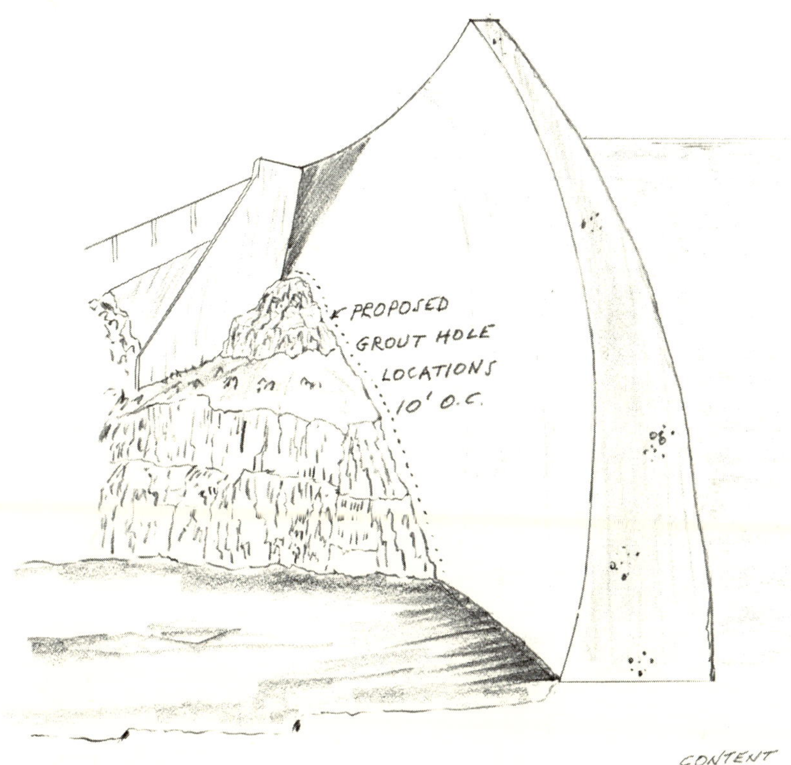

PROPOSED
GROUT HOLE
LOCATIONS
10' O.C.
CONTENT
7-70

SUBJECT: SPECIAL LOCAL GROUTING AT CONCRETE GRAVITY DAM.

CONDITION: ABUTMENT AND FOUNDATION ROCKS ARE WELL-STRATIFIED METAVOLCANICS WHICH DIP 45° OUT OF RIGHT ABUTMENT HILL.

PROBLEM: UPSTREAM-DOWNSTREAM TRENDING BASALTIC DIKE INTRUDES OLDER METAVOLCANIC ROCKS AND IS POTENTIAL PATH OF LEAKAGE BENEATH DAM.

SOLUTION: REMOVE WEATHERED ROCK THEN GROUT ACROSS CONTACT WITH GROUT HOLES 10-FEET ON CENTERS.

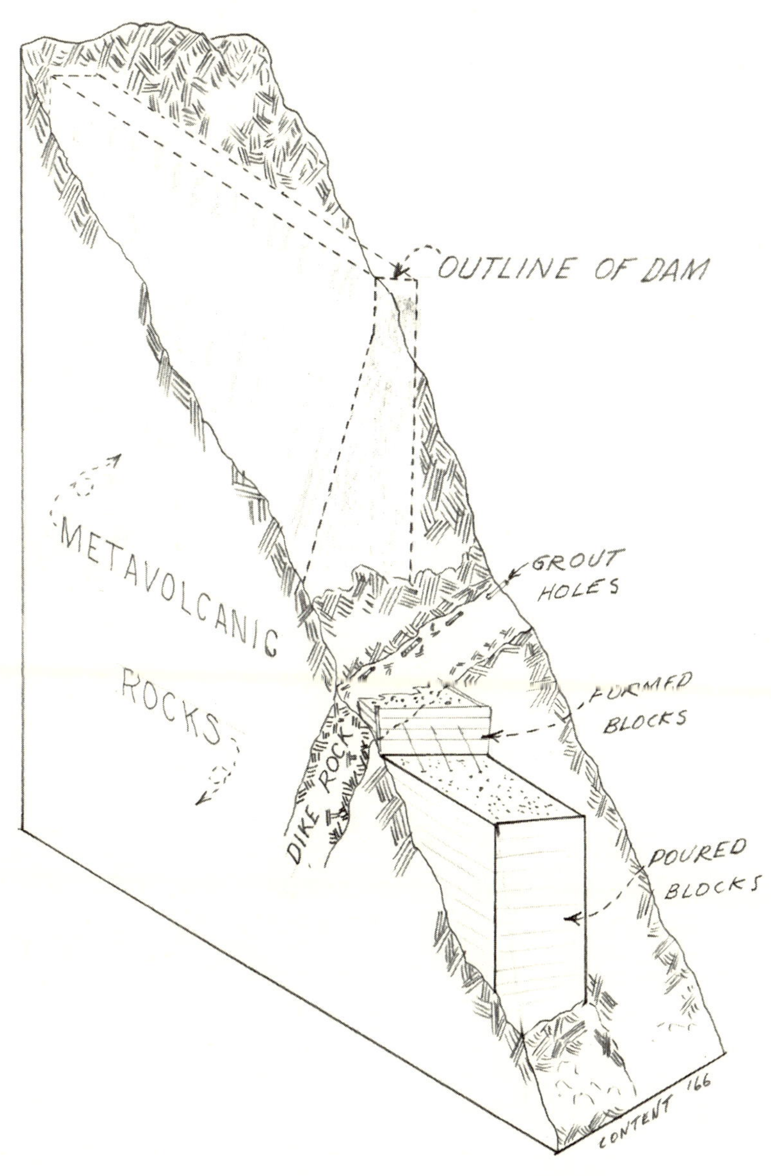
OUTLINE OF DAM
METAVOLCANIC
ROCKS
GROUT HOLES
FORMED BLOCKS
DIKE ROCK
POURED BLOCKS
CONTENT 166

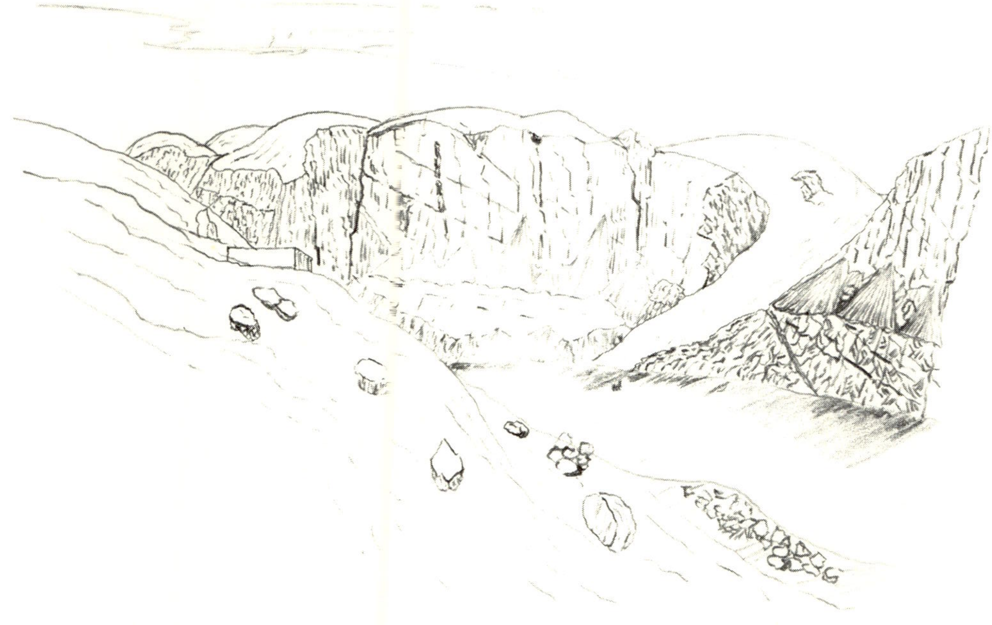

MARMORIK GREENLAND
CONTENT 6-73

<u>SUBJECT</u>: LOW DAM PROPOSED FOR GORGE CUT IN GLACIAL OUTWASH MATERIALS.

<u>CONDITIONS</u>: GLACIAL AND AQUEOGLACIAL MATERIALS UNDER DAM AND RESERVOIR SITES. BOULDER-CHOKED GORGE AT DAM AXIS.

<u>PROBLEM</u>: PERMEABILITY OF BOULDER DEPOSIT. TOO COSTLY TO REMOVE BOULDERS AND PROVIDE POSITIVE CUT-OFF.

<u>SOLUTION</u>: MOVED UPSTREAM TO MORE ACCEPTABLE SITE.

DAM SITE
TERRACE
GLACIAL
MORAINE
MARAROA
RIVER
NEW ZEALAND
3-61
CONTENT

SUBJECT: POSSIBLE INCREASE OF WATER STORAGE IN GLACIAL LAKE.

CONDITIONS: AN EXISTING GLACIAL LAKE IN WYOMING STORES WATER BEHIND A RECESSIONAL MORAINE IN PART CAPPED BY SLIDE MATERIAL.

PROBLEM: RECONNAISSANCE GEOLOGY SUGGESTS WATER BEHIND RECESSIONAL MORAINE DOES NOT RISE ABOVE LIP OF ROCK BASIN GOUGED OUT BY ICE. LIP NOW BURIED IN MORAINE, AND LAKE WATER REACHES MAXIMUM ELEVATION AT CREST OF LIP THEN FLOWS THROUGH COARSE MORAINAL MATERIAL AND SURFACES AS SPRINGS ABOUT ½ MILE DOWNSTREAM. PROJECT PLANS SUGGEST ARRESTING FLOW THROUGH MORAINE TO RAISE HEAD ABOUT 50 FEET. QUESTION IS HOW BEST TO CUT OFF THE FLOW.

SOLUTION: GROUTIG OR UPSTREAM BLANKET HAVE BEEN SUGGESTED; NEITHER HAS BEEN TRIED. PERHAPS SLURRY CUT-OFF TRENCH AT FREEBOARD ELEVATION WOULD BE SUCCESSFUL.

WYOMING 1146
CONTENT

TALUS
SLOPES
TASÜSTÜ DAM SITE AREA
TURKEY 12-66
CONTENT

GLACIER
BOUSER RIVER
BRITISH COLUMBIA
CONTENT

SUBJECT: A DAM SITE

CONDITIONS: GLACIAL LAKEBEDS FOR FOUND-
ATION MATERIALS UNDER LEFT ABUTMENT.
DEEP, OLD ALLUVIUM AND TALUS IN RIVER.
HARD ARGILLITE AND SILICEOUS LIMESTONE
ON RIGHT ABUTMENT.

PROBLEM: WEAK FOUNDATION FOR LEFT
ABUTMENT. DEEP PERMEABLE MATERIALS
UNDER RIVER SECTION.

SOLUTION: VERY FLAT SLOPES FOR DAM
TO ACCOMODATE WEAK LEFT ABUTMENT.
POSITIVE CUT-OFF TO ROCK UNDER ZONE 1.
KEEP ALL APPURTENATE STRUCTURES
ON RIGHT ABUTMENT.

GLACIAL LAKE SILTS
ALLUVIUM & OLD TALUS
BELT FORMATION ROCKS
MONTANA
12-60
CONTENT

SUBJECT: SITE OF OFFSHORE ORE LOADING DOCK.

CONDITIONS: SEAWARD DIPPING BEDS OF GLACIATED MARBLE. DEEP FIORD WATER OFFSHORE.

PROBLEM: EXFOLIATION JOINTS AND BEDDING TEND TO SLAB OFF AT CONSTRUCTION SITE FOR ORE LOADING DOCK.

SOLUTION: GEOLOGIST DIVER EXAMINED ROCK AT PIER FOUNDATION FOOTING SITES. TWO REQUIRED 4½ INCH DIAMETER, 20-FOOT LONG ANCHOR RODS GROUTED INTO ROCK TO PREVENT DOWN DIP MOVEMENT.

WHERE SURFACE SLOPE IS STEEPER
THAN DIP OF STRATA, SLIDES ON
BEDDING ARE LIKELY
BEDDING
JOINT
GREENLAND 1973

SUBJECT: DISCRETE SPIRES OF CONGLOM-
ERATE.

CONDITIONS: VERY LARGE, IN SOME CASES
1×10^6 CUBIC YARDS, SCATTERED BODIES
OF CONGLOMERATE ENCASED IN
SHEARED SHALE.

PROBLEM: ORIGIN OF THESE DISPLACED
ROCKS.

SOLUTION: SOME AUTHORITIES BELIEVE
THESE EXOTICS TO BE OLISTOLITHS. I
THINK THEY ARE SEGMENTS OF A COM-
PETENT CONGLOMERATE STRATUM WHICH
HAVE BEEN TWISTED AND RAFTED OUT
OF POSITION BY BEING SQUEEZED BY
THE LESS COMPETENT SHALE DURING
FOLDING. THE SHALE CONTAINS NUMEROUS
POLISHED PEBBLES WHICH APPEAR TO
HAVE FALLEN FROM THE CONGLOMERATE
PRIOR TO AND DURING FOLDING OF
THE ROCKS.

CONGLOMERATE
"ERRATICS"
SO. TAIWAN 3-74
CONTENT

CONGLOMERATE SPIRES
SOUTH TAIWAN
CONTENT 3-74

KOREAN COUNTRYSIDE
CONTENT 10-69

<u>SUBJECT</u>: DAM SITE IN PERUVIAN ANDES

<u>CONDITIONS</u>: FOLDED LIMESTONE AND
    QUARTZITE AT OR VERY NEAR SUR-
    FACE UNDER PROPOSED DAM AND
    RESERVOIR SITES.

<u>PROBLEM</u>:   PROBABLE CAVERNOUS LIME-
    STONE AND POTENTIAL RESERVOIR
    LOSS ALONG HYDRAULIC GRADIENT
    LINE SHOWN.

<u>SOLUTION</u>:   SEEK NEW SITE AS LEAKAGE
    HERE TOO RISKY AND BLANKETING
    TOO COSTLY.

HYDRAULIC
GRADIENT
PERU 2-65
CONTENT

SUBJECT: STORAGE SITE FOR IMPORTED COOLING WATER.

CONDITIONS: FLAT-LYING SILTSTONES AND SHALES CAPPED BY "CLINKER" BED - A FUSED SHALE. SOMETIMES MISCALLED SCORIA.

PROBLEM: IMPORTED WATER IS PUMPED TO STORAGE SITE; HENCE IS EXPENSIVE. RESERVOIR IS UNDERLAIN BY RELATIVELY IMPERMEABLE MATERIAL BUT BURNED SHALE BED IS PERMEABLE.

SOLUTION: RESTRICT MAXIMUM WATER SURFACE ELEVATION TO BASE OF "CLINKER" BED.

CLINKER
OUTCROP
CONTENT
MONTANA 2-73

BRITISH COLUMBIA
CONTENT

CONTENT
2-68

SUBJECT: SMALL DAM FAILURE

CONDITIONS: THREE-GATED CONCRETE DIVERSION STRUCTURE FOUNDED ON BASALT. LEFT ABUTMENT EARTH WING-DAM ABUTTED INTO VOLCANIC MUDFLOW.

PROBLEM: DURING FLOOD STAGE EXCESSIVE AMOUNT OF TIMBER SLASHINGS FROM LOGGING OPERATIONS SWEPT OVER LOG BOOM JAMMING GATES. DAM WAS OVERTOPPED BREACHING EARTH WING DAM.

SOLUTION: REBUILT STRUCTURE.

PUMICE
VOLCANIC
MUDFLOW
VESICULAR
BASALT
WASHOUT
DIVERSION DAM
ROGUE RIVER. ORE.
1-17-65
CONTENT

SUBJECT: CONCRETE DAM UNDER CON-
STRUCTION.

CONDITIONS: ABUTMENTS OF SOUND
METAMORPHIC ROCKS DIPPING
ABOUT 40° INTO LEFT ABUTMENT.

PROBLEM: FINAL FOUNDATION CLEAN-
UP REVEALED OPEN JOINTS WITH VERY
WELL CONSOLIDATED STREAM GRAVELS
WEDGED INTO THEM. SHOULD GRAVEL
BE REMOVED?

SOLUTION: REMOVED ROCK OVERHANG
BY EXCAVATING ON ½ HORIZONTAL TO
1 VERTICAL SLOPE. REMOVED MOST
OF THE GRAVEL BY HIGH PRESSURE
AIR-WATER JET. SOME LEFT IN PLACE.

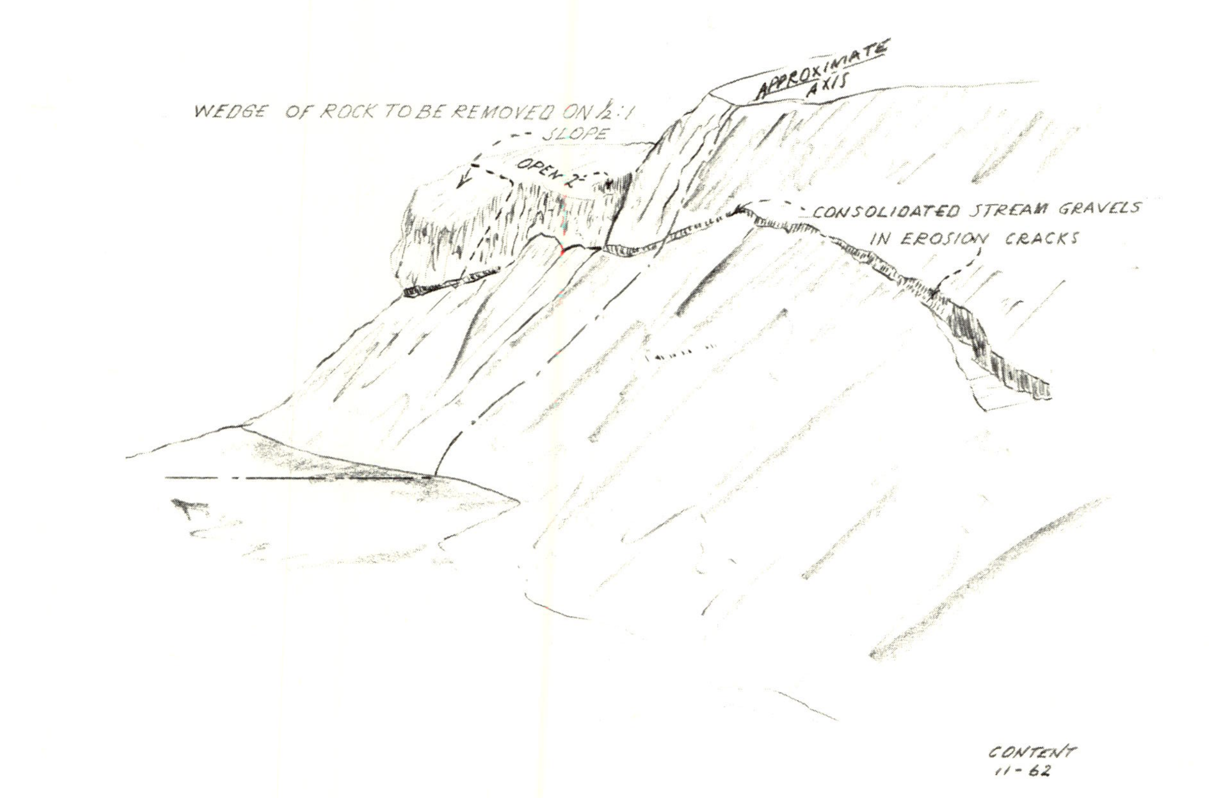

WEDGE OF ROCK TO BE REMOVED ON ½:1 SLOPE
APPROXIMATE AXIS
OPEN 2'
CONSOLIDATED STREAM GRAVELS IN EROSION CRACKS
CONTENT
11-62

SPANISH PEAKS, COLORADO
CONTENT
11-58

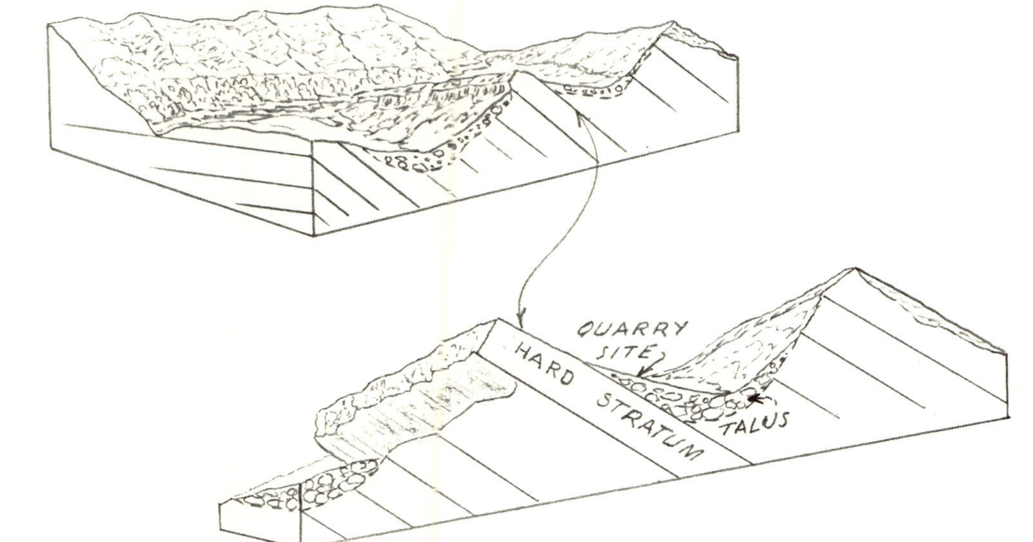

SKETCHES SHOW DIAGRAMMATICALLY
HOW TILTED METAVOLCANIC ROCKS
PROVIDED SUITABLE RIPRAP FROM
HARD STRATUM OF GREENSTONE.

CONTENT
8-61

SUBJECT: LARGE GRANITE MASSES ROUNDED
   BY EXFOLIATION.

CONDITION: THE COARSE-GRAINED
   GRANITE HAS WEATHERED INTO
   IMPRESSIVELY LARGE AND INTER-
   ESTINGLY-SHAPED MASSES OF ROCK.

PROBLEM: NONE - JUST ITEMS OF
   INTEREST.

LOST CREEK AREA, COLO.
CONTENT
8-63

LOST CREEK, COLO.
8-63   CONTENT

BRITISH COLUMBIA

CONTENT

SUBJECT: DAM SITE INVESTIGATION.

CONDITIONS: EXPOSED ROCK IS JOINTED BIOTITE GRANITE.

PROBLEM: SUBSURFACE EXPLORATIONS INDICATED JOINT ARRANGEMENT WAS SUCH THAT BLOCKS OF ROCK ON RIGHT SIDE OF RIVER HAVE MOVED VALLEYWARD ALONG LOW DIPPING SET OF JOINTS. THIS PULLING AWAY FROM ADJACENT ROCK LEFT VOIDS AT VERTICAL JOINT CONTACT.

SOLUTION: SITE WAS ABANDONED.

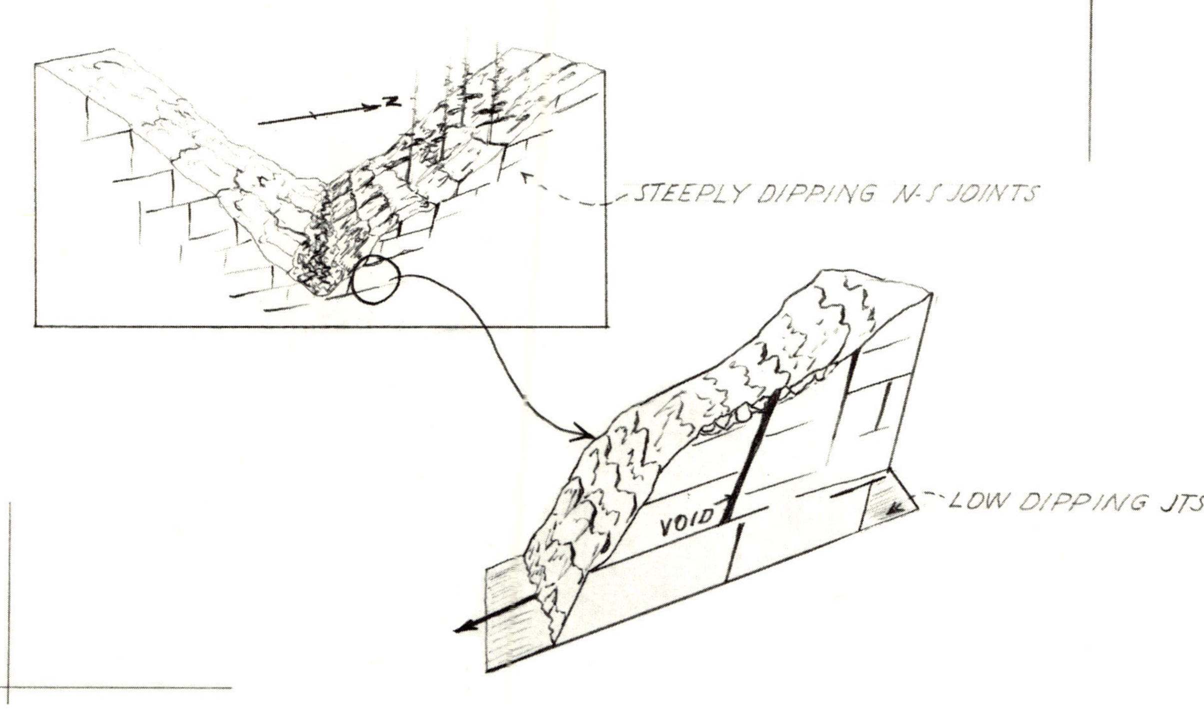

STEEPLY DIPPING N-S JOINTS
N
LOW DIPPING JTS.
VOID
CONTENT
12-59

SUBJECT: DIVERSION CANAL IN
BASALT GORGE.

CONDITIONS: RUN-OF-THE RIVER CANAL
LOCATED ALONG TOE OF SLOPE
ABOVE RIVER.

PROBLEM: RAPID DOWN CUTTING BY
RIVER AND SAPPING OF CANYON
SLOPES BY INFILTRATING WATER
FROM UPLAND CAUSE DOWNDROP
OF LARGE BASALT BLOCKS. THESE
THREATEN CANAL.

SOLUTION: ABANDON PROJECT OR
ACCEPT CONTINUED ABOVE NORMAL
MAINTENANCE.

Iskut toKaniska
B.C.

LAKE MANAPOURI
NEW ZEALAND 4-65
CONTENT

<u>SUBJECT</u>: PROPOSED LOCATION FOR LARGE
STORAGE TANK.

<u>CONDITIONS</u>: FOUNDATION IN AREA OF
MODERATELY WEATHERED GNEISSOID
GRANITE.

<u>PROBLEM</u>: PLANNED LOCATION PLACES
PART OF TANK ON FILL RESTING ON
WEATHERED ROCK THUS SUBJECT TO
FAILURE.

<u>SOLUTION</u>: RELOCATION OF TANK SO
IT IS ENTIRELY IN ROCK CUT. CUT
WAS LINE DRILLED VERTICALLY.

FILL
CUT
CONTENT
11-'64

UPPER TASAJERAS
PERU 9-72
CONTENT